AN OVERVIEW OF SAFETY AND HEALTH FOR WORKERS IN THE HORSE-RACING INDUSTRY

DEPARTMENT OF HEALTH AND HUMAN SERVICES
Centers for Disease Control and Prevention
National Institute for Occupational Safety and Health

An Overview of Safety and Health for Workers in the Horse-Racing Industry

Kitty J. Hendricks, Amia Downes, John Gibbins, Virgil Casini, Elena Page

DEPARTMENT OF HEALTH AND HUMAN SERVICES
Centers for Disease Control and Prevention
National Institute for Occupational Safety and Health

This document is in the public domain and may be freely copied or reprinted.

Disclaimer

Mention of any company or product does not constitute endorsement by the National Institute for Occupational Safety and Health (NIOSH). In addition, citations to Web sites external to NIOSH do not constitute NIOSH endorsement of the sponsoring organizations or their programs or products. Furthermore, NIOSH is not responsible for the content of these Web sites. All Web addresses referenced in this document were accessible as of the publication date.

Ordering Information

To receive documents or other information about occupational safety and health topics, contact NIOSH at

Telephone: **1–800–CDC–INFO** (1–800–232–4636)
TTY: 1–888–232–6348
E-mail: cdcinfo@cdc.gov

or visit the NIOSH Web site at **www.cdc.gov/niosh**.

For a monthly update on news at NIOSH, subscribe to *NIOSH eNews* by visiting **www.cdc.gov/niosh/eNews**.

DHHS (NIOSH) Publication No. 2009–128

April 2009

SAFER • HEALTHIER • PEOPLE™

Contents

Introduction 1

Background of the Horse-Racing Industry in the United States 1

Public Input 2

Safety Issues 3

Data Analyses of Nonfatal and Fatal Injuries 4
    Nonfatal Injuries 4
    Fatal Injuries 4
    Case Report 6

Adverse Health Effects 8
    Musculoskeletal Disorders 8
    Weight Reduction 8
    Lead Exposure 8
    Respiratory Issues 9

Regulations 9
    Weight and Fitness Requirement 9
    Personal Protective Equipment (PPE) 10
    Licensure 12
    Race Track Design 12
    Insurance 13

Conclusions and Recommendations 13
    Industry Representatives (Race Tracks, Racing Commissions
        and Horse Owners) 14
    Jockeys 15
    Professional Associations 15
    Other Race Track Workers 15

References 15

Acknowledgments

The authors are indebted to the staffs at the Randall 'Doc' James Racetrack, the Keeneland Race Course, and the North American Racing Academy for their assistance in facilitating site visits and investigations. Additionally, we appreciate the efforts of the Bureau of Labor Statistics and the Consumer Product Safety Commission for their reviews of the data.

Introduction

The safety and health hazards associated with the horse-racing industry, along with a lack of adequate disability and health insurance for its workers, prompted an investigation by Congress which culminated with hearings in 2005. One of the outcomes from these Congressional hearings was a letter from the Chairman and Ranking member of the Subcommittee on Oversight and Investigations of the U.S. House of Representatives Committee on Energy and Commerce to the Department of Health and Human Services Secretary, requesting assistance from the National Institute for Occupational Safety and Health (NIOSH) in investigating the safety and health hazards in the horse-racing industry.

In response to this request, NIOSH conducted a review of the available safety and health literature on thoroughbred and standardbred horse racing; conducted site visits to two racetracks in Lexington, Kentucky, Keeneland Race Course and the North American Racing Academy; completed a fatality investigation; conducted analyses of injury data from relevant data sources; reviewed regulations governing the horse-racing industry in the United States and other countries; and held a public meeting in order to garner concerns about the health and safety of workers in the horse-racing industry.

This document is intended for all workers associated with the horse-racing industry, including jockeys, other race track workers, horse and race track owners, and racing commissions. The document summarizes NIOSH's efforts in responding to the Congressional inquiry and provides recommendations for reducing the number of injuries and adverse health effects for workers in the horse-racing industry.

Background of the Horse-Racing Industry in the United States

Horse racing is a popular spectator sport in the United States [Press et al. 1993]. It is an ancient sport, having its origins among the prehistoric nomadic tribesmen of Central Asia around 4500 B.C. The first race track in the United States was established as early as 1665 in Long Island, New York [Parker 1998]. Currently, thoroughbred race tracks exist in more than half of the States, with over 125 tracks in operation [USA Horse Racing 2003].

Little is known about the health status or number and nature of injuries and illnesses that are sustained by workers in the horse-racing industry. Many risk factors are involved when a 115-pound jockey rides an 1,100 pound animal running 40 miles per hour. Besides the jockey, other workers (e.g., backstretch workers, farriers, grooms, trainers, starting gate attendants, etc.) have their own safety and health considerations.

The total number of workers in the horse-racing industry is hard to determine. Employment data from customary sources, such as the Current Population Survey [U.S. Census Bureau 2008], are categorized by broad industry group. The horse-racing industry falls within the broad group of spectator sports, which includes auto racing, football, etc. Since subgroup distributions by sport are not available from the group of spectator sports, other sources for employment data were considered, including the American Horse Council Foundation and the Jockeys' Guild of America.

In 2004, the American Horse Council Foundation commissioned a study on the economic impacts of the horse-racing industry in the United States. This study estimated the equivalent of 146,625 full-time employees directly working in the industry [American Horse Council Foundation 2005]. Directly employed workers include jockeys, trainers, exercise riders, grooms, valets, starting gate attendants, apprentice jockeys, and veterinarians. The breakdown of workers into each of these categories was not reported in the Council's study.

The Jockeys' Guild of America represented an estimated 1,200 riders nationwide in 2007 [Jockeys' Guild 2007]. To qualify as a member of the Guild, a jockey must hold a valid, unrevoked jockey's license. A statistical study of jockeys' mounts for 2005 reported 1,908 licensed professional jockeys rode during that year [Colton 2007]. However, not all licensed jockeys are members of the Guild.

The horse-racing industry presents a demanding lifestyle. Most work days start at 4 a.m. and often continue late into the night. Workers travel from track to track, and traveling introduces another risk. Commonly, jockeys compete in more than 1,000 races a year, often riding several hundred different horses [Burwinkle 2002].

Jockeys are considered independent contractors [Gitomer 2005], as are many employees associated with horse racing; they may not be covered by the Occupational Safety and Health Administration (OSHA) or by the Department of Labor's Wage and Hour Division. Opacich and Lizer [2007] report that backstretch workers, considered independently contracted agricultural workers, and jockeys are both exempt from minimum wage requirements and often are not entitled to workers' compensation or social security benefits.

The safety and health concerns associated with horse racing are numerous. While health issues, particularly those associated with weight reduction, have become more recognized, there is a lack of scientific literature concerning safe work practices and the use of proper personal protective equipment (PPE). Furthermore, as new technologies such as synthetic tracks become more common, the impact they may have on worker safety and health must be addressed.

Public Input

On May 22, 2007 NIOSH held a public meeting to garner concerns about the health and safety of workers in the horse-racing industry. The meeting, "Safety and Health in the Horse Racing Industry and Best Practices," was attended by 26 individuals, representing 16 different agencies, including the Jockeys' Guild, the National Thoroughbred Racing Association, the American Horse Council, the Grayson-Jockey Club Research Foundation, and the Racing Medication and Testing Consortium. In addition to the meeting, presentations, and ensuing discussions, a docket was established to receive comments from the public [NIOSH 2007a].

NIOSH received nine submissions to the docket covering a range of topics. Among these were comments on the many health issues that jockeys face. Many of these health effects are related to weight-reduction and weight-maintenance practices. Other areas of concern included exposure to lead, and the effects of repeated head trauma. Other submissions included information on the use of PPE, requirements and qualifications for on-site emergency medical service at race tracks, the health disparities found between those employed in the horse-racing industry and the general population, barn fires, new technologies in racing surfaces, and other environmental health issues. Although many of these submissions were anecdotal, two submissions described research-in-progress from academic institutions.

Safety Issues

The Jockeys' Guild of America reported that over 100 jockeys have been killed from 1950 through the mid 1980s [DeBenedette 1987], and the focus of horse-racing injuries has primarily been on jockeys [Waller et al. 2000; Turner 2000; Press et al. 1993]. However, others employed in the horse-racing industry are exposed to many of the same safety risks as jockeys. In fact, Turner [2000] indicates that many injuries occur during morning warm-ups, which are more likely to involve trainers or exercise riders.

An analysis of injuries to licensed jockeys by Waller et al. [2000] identified 6,546 injuries (606 injuries/1,000 jockey years) and three fatalities from 1993 through 1996. This study also found that 44% of the injuries resulted from the jockey being thrown from the horse, with the head, neck, and face incurring the most injuries (19%). A survey completed by 706 professional jockeys collected information on the number and types of injuries that they incurred throughout their careers [Press et al. 1993]. These jockeys reported 1,757 total injuries, with fractures being the most common.

Safety in the horse-racing industry is a complex subject. There are obvious hazards associated with riding a racehorse, and there are other hazards that may be associated with the track itself. Furthermore, hazards are often associated with off-track horse activities.

PPE in this industry has undergone considerable change in recent years. Standards associated with helmets and protective vests are now regulated in several states. Also, engineering controls have been implemented in the industry including padded starting gates, new safety rails along the track to absorb much of an impact should a jockey be thrown against them, and new track surfaces intended to make a safer, more consistent racing surface.

Waller [2000] found that the start gate is one of the most common sites for injury events. The starting gate contains a horse and mounted jockey in a small, restricted area, presenting an opportunity for the jockey to be crushed against a rigid surface by the horse.

New track surfaces, especially synthetic track surfaces, have gained popularity in recent years and have been installed on some of the nation's premier race tracks. Although each of the available brands of synthetic track varies in composition, all contain some combination of synthetic fibers mixed with sand. These synthetic surfaces are designed to have a cushioning effect meant to reduce the risk of injury for horses and to maintain a more consistent racing surface. However, the synthetic fibers from these surfaces may present an inhalation risk for workers in the industry. Although synthetic surfaces claim to reduce catastrophic injuries to horses and jockeys, no quantitative assessments or peer-reviewed published data are available to substantiate these claims.

Safety reins are another area where improvements have been made. When reins snap during a race, the injuries resulting from the loss of control of a horse can often be very severe. Implementation of safety reins, which is a type of rein that is reinforced with a wire cable, nylon strap, or other safety device attached to the bit, is a simple solution to broken reins and the subsequent loss of control of the horse. Some tracks have already instituted their use.

Data Analyses of Nonfatal and Fatal Injuries

Nonfatal Injuries

NIOSH reviewed nonfatal, emergency-department data from the National Electronic Injury Surveillance System occupational supplement (NEISS-Work). NEISS data are collected by the Consumer Product Safety Commission (CPSC), which shares them with NIOSH through an interagency agreement. NEISS-Work provides nationally representative data for persons treated for nonfatal work-related injuries and illnesses in U.S. hospital emergency departments.

NEISS-Work uses a nationally stratified probability sample of 67 U.S. hospitals with 24-hour emergency departments (EDs).* Hospitals in the sample were selected from the approximately 5,300 rural and urban U.S. hospitals after stratification into four size-based strata (i.e., by total annual ED visits) plus a children's hospital stratum. Each injury/illness was assigned a statistical weight correlating to the probability of selecting the treating hospital within its sample stratum. Weights were adjusted monthly for nonresponse among the sample hospitals and annually for national fluctuations in ED use. NIOSH estimates that ED-treated injuries and illnesses in NEISS account for approximately one-third of all U.S. work-related injuries and illnesses that require medical treatment [CDC 2001].

On the basis of a NIOSH review of NEISS-Work case narratives,† an estimated 14,200 injuries and illnesses (95% confidence interval [CI] 8,400 to 19,900) associated with the thoroughbred and standardbred horse-racing industry occurred in the United States from 1998 through 2006. Males incurred 61% (8,700, 95% CI 5,000 to 12,400) of the injuries compared to 39% (5,500, 95% CI 2,600 to 8,400) for females. The highest proportion of injuries (33%) occurred to workers from 35 years to 44 years of age. Table 1 presents a complete breakdown of injuries by age.

The part of the body injured most often, at 33%, was the upper extremities (i.e., the shoulder, upper arm, elbow, lower arm, wrist, hand, and fingers). A distribution of injured body parts appears in Table 2. In almost all of the cases of nonfatal injuries (91%), the injured person was treated and released, with the remainder either being admitted or transferred to another hospital. When looking at what times of year nonfatal occupational injuries occur, 67% (9,491, 95% CI 5,500 to 13,482) occurred from April through September.

Fatal Injuries

The Census of Fatal Occupational Injuries (CFOI) is a multi-source data system maintained by the U.S. Bureau of Labor Statistics (BLS) to identify work-related deaths in the United States. To better meet the research needs of NIOSH, BLS provided NIOSH with a detailed research file that includes variables such as specific age (NIOSH research was conducted with restricted access to BLS data. The views expressed here do not necessarily reflect the views of the BLS.) The New York City Department of Health did not authorize the release of these more detailed data to NIOSH; therefore, data from New York City are excluded from these analyses. A NIOSH review

---

*The NEISS-Work data collection system is operated by the Consumer Product Safety Commission (CPSC) as a supplement to its NEISS surveillance of consumer product-related injuries. NEISS-Work estimates include all work-related injuries regardless of product involvement. NEISS-Work uses approximately two-thirds of the CPSC sample of 101 hospitals. Because of hospital closures and other nonparticipation/nonresponse factors, the number of reporting hospitals can vary monthly and yearly.

†Cases were identified by conducting a narrative search for the terms "trainer," "exercise rider," "horse," "race," "groom," "jockey," "starter," "hot walker," "outrider," and "valet."

Table 1. National estimates of nonfatal work-related injuries in the horse-racing industry presented to hospital emergency departments by age group, 1998–2006

Age group	Nonfatal Injuries	Percentage	95% Confidence interval*
Less than 25 years	3,000	21	1,000 to 5,000
25–34 years	3,300	23	1,600 to 4,900
35–44 years	4,600	33	2,300 to 6,900
45+ years	3,300	23	1,800 to 4,700
Total	14,200	100%	8,400 to 19,900

Source: National Electronic Injury Surveillance System—Occupational Supplement.
*Confidence intervals may not be symmetrical due to rounding.

Table 2. National estimates of nonfatal work-related injuries in the horse-racing industry presented to hospital emergency departments by body part injured, 1998–2006

Body part injured	Nonfatal injuries	Percentage	95% Confidence interval*
Upper extremities	4,700	33	2,400 to 7,000
Lower extremities	3,600	25	1,700 to 5,500
Head/neck	3,200	23	1,900 to 4,500
Abdomen/trunk	2,600	18	1,400 to 3,700
Internal injuries	NR†	NR†	—
Multiple body parts	NR†	NR†	—
Total	**14,200**	**100%**	**8,400 to 19,900**

Source: National Electronic Injury Surveillance System—Occupational Supplement.
*Confidence intervals may not be symmetrical due to rounding.
†Estimate is not reportable or is suppressed because of a non-reportable cell.

Table 3. Work-related fatalities in the horse-racing industry by occupation, U.S., 1992–2006

Occupation*	Number of deaths	Percentage
Trainer	28	35
Jockey	26	33
Exercise rider	8	10
Groom	7	9
Other track personnel†	10	13
Total	**79**	**100%**

Source: Census of Fatal Occupational Injuries, 1992–2006. Special research file provided to NIOSH; excludes New York City decedents.
*Occupation was coded through a combination of the occupation code and the occupation narrative fields.
†Includes owners, farriers, stable workers, handlers, managers, etc.

of CFOI narratives along with industry and occupation variables[‡] identified 79 deaths between 1992 and 2006 associated with the thoroughbred and standardbred horse-racing industry, resulting in an annual average of 5.6 fatalities per year (Figure 1). The level of detail available in the narrative information on which the identification of these deaths is based varies. Because of this, 79 is likely an underestimate of the true number of occupational fatalities in the horse-racing industry.

Males accounted for a majority (65, 82%) of the 79 fatalities between 1992 and 2006. Twenty-one percent of the decedents were Hispanic, which is higher than the proportion of Hispanic occupational fatalities for all industries over the same time period (12.5%) [BLS 2008]. When examining these deaths by occupation, trainers (35%) and jockeys (33%) incurred the majority of the fatalities (Table 3). The age of the decedent ranged from less than 20 years to greater than 65 years, with an average age of 45 years. Table 4 provides information on fatalities by age group.

A further breakdown of fatalities and exact age range could not be provided due to confidentiality requirements.

Forty-eight percent (38) of the fatalities were sustained on the grounds of the race track, but not during an actual race. Twenty-three percent (18) of the deaths occurred during a race. In 49% of the fatalities, the decedent was either thrown or fell from or with the horse. Seventeen (22%) of the fatalities occurred as a result of either being kicked or stepped on by the horse. Of these 17 deaths, 14 resulted from a kick to the chest or abdomen. The narratives in these cases did not indicate if PPE was worn at the time of injury.

When looking at what times of year fatalities occur, 62% (49) of deaths in the horse-racing industry occurred from April through September. Figure 1 shows the number of deaths by year.

Case Report

The following case report is an example of several risks that jockeys in the horse-racing industry face while at work. This case was investigated

---
‡Cases were identified by conducting a narrative search for the terms "trainer," "exercise rider," "horse," "race," "groom," "jockey," "starter," "hot walker," "outrider," and "valet."

Table 4. Work-related fatalities in the horse-racing industry by age group, U.S., 1992–2006

Age group	Number of deaths	Percentage
< 20 years	6	8
20-24 years	5	6
25-34 years	11	14
35-44 years	19	24
45-54 years	14	18
55-64 years	11	14
> 65 years	13	16
Total	79	100%

Source: Census of Fatal Occupational Injuries, 1992–2006. Special research file provided to NIOSH; excludes New York City decedents.

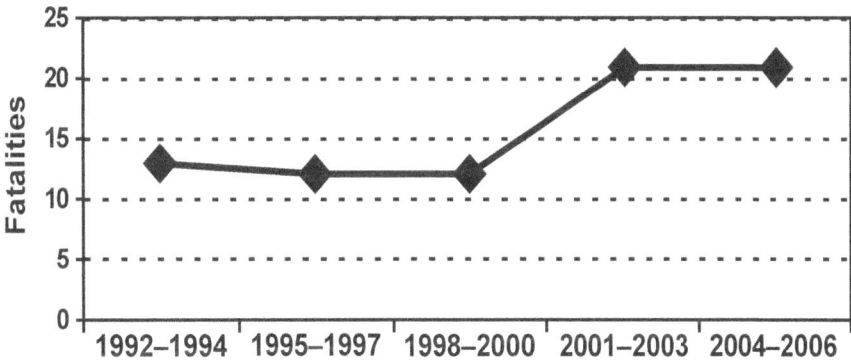

Figure 1. Work-related fatalities in the horse-racing industry by year, U.S., 1992–2006.

Source: Census of Fatal Occupational Injuries, 1992–2006. Special research file provided to NIOSH; excludes New York City decedents.

by NIOSH through the Fatality Assessment and Control Evaluation Program [NIOSH 2007b].

On February 18, 2007, a 65-year-old male jockey died after being thrown from his mount in the starting gate during the start of a race. The jockey and his mount were being led into the eighth stall of an eight-stall mobile starting gate. As this was being done, the horse in the fifth stall reared and struck the door of the starting gate. All eight stall gates opened simultaneously, effectively starting the race. Whether the starting gate had opened because of the horse bolting in the fifth stall or because of operator error is not clear. A videotape review of the race indicated that, at the time of the gates opening, the jockey in the eighth stall had not completely settled onto his mount. The jockey was thrown backward and appeared to hit his head on the back door of the stall. Since the horses would have to pass the location of the starting gate to complete the race, the starting gate was hooked to a tractor for quick removal from the track. When the tractor operator saw the horses break from the starting gate, he began to pull the starting gate from the track. When other track workers saw the horse without a rider, they signaled for the tractor operator to stop. The starting gate, which had been pulled up onto the abdomen of the thrown jockey, was backed off the victim by the tractor operator. Rescue personnel stationed in an ambulance at the track responded immediately. The victim was unconscious. Three additional ambulances and a rescue unit responded within minutes. He was transported to the hospital where he died in surgery approximately four hours after the incident. The victim had suffered broken ribs and a ruptured spleen. The death certificate stated that the victim died of complications during surgery to repair a damaged spleen.

Adverse Health Effects

Adverse health effects is another area of research that has been focused primarily on jockeys. Although eating disorders and their long- and short-term effects have been widely studied in this population [Leydon and Wall 2002; Labadoarios et al. 1993; Bishop and Deans 1996; King and Mezey 1987; Price 1973], other health effects, such as musculoskeletal concerns [Tsirikos et al. 2001; Lavelle and Murphy 1977] have also been studied.

Musculoskeletal Disorders

Rather than riding directly in the saddle, jockeys use their legs for gripping, stability, and balance. This forces them into a forward lean, creating a forced static posture over the horse. Jockeys are subject to dynamic and static joint loading, impact loading, and injuries associated with acceleration and deceleration from racing. The combination of the applied forces, static postures, repetitive motion, and trauma from joint loading may lead to musculoskeletal disorders of the lower extremities and spine [Tsirikos et al. 2001; Lavelle and Murphy].

Weight Reduction

The competitive nature of this sport, added to existing pressures for jockeys to maintain low body weights, increase the risk for jockeys to acquire disordered eating habits and to adopt other unhealthy behaviors [Leydon and Wall 2002; Labadoarios et al. 1993; Bishop and Deans 1996; King and Mezey 1987; Price 1973]. Examples include vomiting, abusing laxatives and diuretics, using saunas and hot baths to lose water weight, exercising excessively, smoking to curb appetite, restricting or avoiding food, taking diet pills, and restricting fluid intake. These disordered eating habits and other weight-loss behaviors can result in short- and long-term health effects such as dental erosion, nutritional deficiencies, menstrual irregularity, low bone density, dehydration, and heat stress. Cardiac arrhythmias have been reported in individuals with anorexia nervosa and bulimia nervosa [Mitchell and Crow 2006; Palla and Litt 1988]. Renal abnormalities, including kidney stones and renal failure, have also been reported in these populations [Jonat and Birmingham 2003; Inui et al. 1997; Copeland 1994; Palla and Litt 1988]. However, the medical literature does not report any studies of cardiac arrhythmias nor renal abnormalities specifically for jockeys.

Lead Exposure

Historically, uncovered lead plates have been added to saddles to adjust the weight each horse carries as determined by handicapping for races that use a pari-mutuel betting system.§ The lead plates are typically handled by the valets and/or the clerks of scales, but some jockeys handle them as well. The plates are often stored in an open box in the jockey room. When plates are thrown into the box for storage, lead dust is generated. A number of tracks have eliminated the use of uncovered lead weights, either encasing lead weight in leather or another type of cover or using weighted pads that are placed under the saddle. There are no published studies of blood lead levels in jockeys or environmental lead levels in jockey rooms. Exposure to lead in dust occurs primarily through ingestion from hand-to-mouth contact. Since gastrointestinal absorption of lead is lower in adults than in children, ingestion of lead by these workers does not pose the same risk as it does for children [Diamond et al. 1998].

Respiratory Issues

Several race tracks across the country recently began using a composite synthetic track surface. Synthetic surfaces, which have been in use in Europe for over two decades, are a combination of polypropylene fibers, recycled rubber, and silica sand covered in a wax coating. Little information is available about how this material deteriorates over time from rain, sunlight, heavy use, and other track conditions. Potential health concerns are the effects due to the release of respirable silica over time [LaMarra 2007]. There has been no research comparing the two surface types in regard to the safety and health of the jockey.

---

§Pari-mutuel betting is a betting system in which all bets of a particular type are placed together in a pool; taxes and a house "take" or "vig" are removed, and payoff odds are calculated by sharing the pool among all placed bets.

Regulations

Rules and regulations in the horse-racing industry are determined by individual states and can be referenced through sources such as state racing commissions, administrative codes, horse-racing boards, and business regulations. There are no national standards or regulations for this industry.

This document summarizes regulations from nine horse racing states. Five of these states, including California (CA), Illinois (IL), New York (NY), Ohio (OH) and Kentucky (KY), were selected based on their recognition within the horse racing industry [California Horse Racing Board 2007; Joint Committee on Administrative Rules 1994; New York State Legislature 2007; Ohio State Racing Commission n.d.; Kentucky Administrative Regulations 2007]. The other four states, Delaware (DE), Washington (WA), West Virginia (WV) and Pennsylvania (PA), were chosen because of some unique feature in their respective state's rules and regulations governing horse racing [Delaware General Assembly 2007; Washington State Legislature 2007; West Virginia Racing Commission 2000; Pennsylvania Code 1997].

Weight and Fitness Requirement

Of the states reviewed, only California imposes a maximum weight limit (125 lbs.) for jockeys to be licensed. Four states (IL, PA, WV, and CA) require annual physical examinations for all jockeys. In California and Illinois, these must be conducted by a physician who has been approved by the state's Horse Racing Board. Six states (DE, KY, NY, OH, WA, and WV) can also require proof of fitness to ride before any race, if the board or stewards have reasonable concerns about a jockey's health. Proof of fitness to ride includes a physical examination and a medical affidavit signed by a physician prior to the start of the race and can also include tests for drug or alcohol use.

Personal Protective Equipment (PPE)

Helmets

Regulations from all nine states had at least a minimum requirement regarding helmets. The least restrictive state (CA) requires only jockeys to wear a properly fastened helmet. Five other states (DE, IL, NY, OH, and PA) require the helmet be approved by the Stewards¶ or the state racing commission. The most stringent states (KY, WA, and WV) require helmets meet the American Society for Testing and Materials (ASTM) standard F1163-04 [ASTM 2007a], "Specifications for Headgear Used in Horse Sports and Horseback Riding."

Personal protective equipment (PPE) rules also vary by state in regards to who is required to wear a helmet and when. While a few states require anyone mounted on horseback to wear a helmet (NY, PA, and WA), Illinois requires helmets to be worn by jockeys and stable employees only [Joint Committee on Administrative Rules 1994]. Ohio requires helmet use for jockeys while racing but also stipulates that anyone working out a horse on a flat racing strip must wear a helmet. Kentucky and Delaware have no written requirements about when a helmet must be worn. Table 5 details, by state, the standards established for PPE.

Vests

Five of the six states that require jockeys to wear safety vests also specify that the vest have a shock absorption protection rating of at least 5 as certified by the British Equestrian Trade Association (BETA) (IL, KY, NY, OH and WV) [BETA 2007]. California requires that the vest meet additional standards in regards to body coverage [California Horse Racing Board 2001], and Washington requires that vests meet the ASTM/Safety Equipment Institute (ASTM/SEI) standard F1937-04, "Specifications for Body Protectors Used in Horse Sports and Horseback Riding" [ASTM 2007b; Washington State Legislature 2007].

Requirements among states for the use of a safety vest have even greater variation than those for wearing a helmet. West Virginia requires vests only for jockeys, and Washington has rules requiring vests for all persons on horseback. Ohio and West Virginia require that the equipment must be worn only when racing. Neither Delaware nor Pennsylvania specify who must wear a vest or when it must be worn, but California, Kentucky, and New York have established rules requiring a vest essentially at all times when mounted on horseback (racing, training, exercising, warming up, or parading). Table 5 outlines the specifics of PPE use.

Safety Reins

In addition to these two pieces of PPE, Ohio has adopted a rule mandating the use of safety reins; this rule went into effect July 1, 2008. The rule states that no person mounted on horseback that is riding, breezing, exercising, galloping, or working out the horse on facility grounds under the jurisdiction of the commission will do so without using safety reins [Ohio State Racing Commission n.d.].

PPE Weight Requirements

The only aspect of PPE use that does not appear to have much variation from state to state is related to how the equipment affects a jockey's

---

¶ To be accredited as a Steward for flat racing, an applicant must complete a 60-hour educational seminar, pass an examination, and meet the requirement for amount of experience in the racing industry [Racing Officials Accreditation Program 2007].

Table 5. Safety equipment regulations* in the horse-racing industry for selected states

Personal Protective Equipment	CA	DE	IL	KY	NY	OH	PA	WA	WV
Helmet									
Properly Fastened	•								
Approved by Commission, Board or Stewards		•	•		•	•	•		
ASTM F1163-04a				•				•	•
Who Must Wear a Helmet									
Jockey		•	•						•
Apprentice Jockey, Stable Employees			•						
Any Person on Horse on Flat Racing Strip	•			•	•	•	•		
When Helmet Must be Worn									
Not Specified		•	•						
Racing, Exercising	•			•	•	•	•		•
All Mounted Persons on Facility/ Association Grounds								•	
Type of Vest									
No Requirement							•		
Approved by Commission/Board		•							
BETA Vest (rating of 5)			•	•	•	•			•
BETA Vest (rating of 1), ASTM F1937-04								•	
BETA Vest (rating of 5), Coverage Requirement	•								
Who Must Wear a Vest									
Not Specified		•		•			•		
Jockey	•		•		•	•			•
Apprentice Jockey						•			
Exercise Rider, Stable Employee	•		•		•				
All Mounted Persons								•	
When Vest Must be Worn									
Not Specified		•					•	•	
Racing	•		•	•	•	•			•
Schooling, Exercising	•		•	•	•				
Training, Warm-up, Parading	•			•	•				

*State regulations are subject to change. These regulations are accurate as of January 2008.
BETA=British Equestrian Trade Association
ASTM=American Society for Testing Materials

weight prior to a race. States appear consistent in that all safety vests can weigh no more than two pounds and the weight of PPE (e.g., helmets and protective vests) should not be included in the jockey's weight when weighing in. There appears to be very little variation in the weight of the PPE used by the majority of jockeys.

PPE Requirements in Other Countries

Jockeys in other countries, such as Australia and Ireland, are subject to more stringent rules and regulations related to safety equipment than even the most strict states in the United States. For example, with respect to helmets, these two countries have rules specifying serviceable condition, types of acceptable chin straps, correct sizing, requirements for helmet liners, and the use of a mounted safety warning light on helmets when worn in darkness [Australian Racing Board 2007; The Turf Club 2007]. Ireland has taken additional steps to ensure safety by recommending that all riders wear safety goggles [The Turf Club 2007].

Licensure

In the United States, jockey licenses are handled by state agencies; there are no national requirements for jockey licensure. For the states that were reviewed, there are only two licensure requirements common to every state: a minimum age limit and evidence of physical fitness. However, the minimum age limit and how physical fitness is assessed are not consistent among states. In most states (CA, DE, KY, NY, PA, and WV), jockey applicants must be at least 16 years of age. Ohio and Washington require their applicants to be at least 18 years of age. Evidence of physical fitness to ride varies from providing a medical affidavit or equivalent stating the jockey is fit to ride (DE, KY, NY, OH, and WA) to an annual physical examination (IL, PA, WV, and CA), with California also specifying that the exam must include vision and hearing screenings. Depending on the state, these examinations are either administered by a licensed physician of the jockey's choice (OH, PA, and WV) or a licensed physician approved or designated by the Racing Commission, Board or Stewards (CA and IL). Overall, the requirements for obtaining a license have few health or safety implications.

Additional requirements include maximum weight limits (125 lbs. in CA) and requirements for the minimum number of races an applicant has to have ridden in before being granted a jockey's license (two in DE and three in KY). California requires applicants to pass an examination of the rules and regulations or to demonstrate their qualification through some other assessment. Some states (NY, PA, and WV) stipulate the number of races one can ride in as an apprentice jockey or temporary permit holder before attaining a jockey's license, which ranges from 2 to 10 races. Applicants in Delaware and Kentucky must have served in the stables 1 year prior to applying for their license.

Race Track Design

Several of the states that were reviewed have regulations pertaining to the design of race tracks. New York and Pennsylvania have a general rule in regards to the safety of racing facilities in their respective states. These states charge each racing association with maintaining racing facilities in good condition and having tools and equipment available to maintain a uniform track, weather permitting.

California has some of the most stringent rules related to race-track safety. California regulations include specifications for rails (permanent vs. non-permanent), rail posts, turf course paths, distance between rails and other objects, drainage ditches, rail gate openings, ingress and egress gates or gaps, lighting systems, and track surface elevation [California Horse Racing Board 2006a]. In 2006, the California Horse Racing Board approved a rule requiring the installation of a polymer, synthetic-type racing surface for all tracks that operate 4 weeks of continuous thoroughbred racing in a year. All tracks were to comply with this rule by December 31, 2007 [California Horse Racing Board 2006b].

Insurance

One of the objectives of modern workers' compensation law is the encouragement of safe and healthful workplaces [National Commission on State Workmen's Compensation Laws 1972]. Jockeys are independent contractors and therefore do not qualify for workers' compensation under their respective states' statutes. A limited number of states have established funds specifically for jockeys to address this problem.

Delaware has established a Jockey's Health and Welfare Board to administer the Jockey's Health and Welfare Benefit Fund, which is used to provide health coverage to active jockeys who regularly ride in Delaware, eligible retired jockeys, and disabled Delaware jockeys [Delaware General Assembly 2006]. The fund also pays for health coverage for eligible dependents of these groups of jockeys. Money from licensed video lottery agents and the purse account are transferred and maintained in an account at the state's Department of Agriculture. Through investments made by the state Treasurer, proceeds from this account are used by the Fund to pay for health coverage.

New York has established a Jockey Injury Compensation Fund [New York State Legislature 2007]. These funds purchase workers' compensation insurance coverage on a blanket basis for jockeys in their state. All owners and trainers licensed in this state are required to contribute to these funds to cover the cost of the premium. In California, no persons applying to be an owner or trainer will be granted a license unless they have secured workers' compensation insurance for licensed employees and proof of insurance can be submitted to the Board [California Horse Racing Board 1978]. If coverage is cancelled or terminated, the trainer's or owner's license will be automatically suspended with grounds for revocation of the license.

One significant drawback to the funds is the lack of availability of coverage when riding in another state or country. For example, a rider from New York who was permanently disabled while riding in a race in Kentucky would not be eligible to collect New York workers' compensation because the injury was sustained out of state.

Conclusions and Recommendations

Data show that between the years 1998 and 2006 an estimate of more than 14,000 occupational injuries associated with the horse-racing industry were treated in U.S. hospital emergency rooms. Further, between 1992 and 2006, 79 deaths occurred to those working in this industry. These numbers are almost certainly underestimates of the true numbers.

The data demonstrate that jockeys are not the only workers exposed to hazards in this industry. Trainers, grooms, exercise riders, and various others encounter many of the same hazards as jockeys, and it is important that these occupations have the same health and safety protections, for example, wearing safety vests and helmets when in close proximity with horses.

The true risk for injuries in this industry cannot be properly evaluated without sustained data collection over an extended period of time, including collecting data on the number of workers in this industry and the injuries and fatalities that occur. It is not within the scope of current national surveillance systems to collect data to the detail necessary to accurately capture these injuries. Furthermore, without appropriate denominator data, injury rates cannot be calculated. The calculation of injury rates would allow for meaningful comparisons to workers in other industries. In order to accurately collect these data, a standardized injury report form could be created to collect critical information about injury incidents. This information could then be recorded and maintained in a centralized database. The development of such a system would allow for meaningful analyses to determine the etiology of injury in the United States for this industry. Northern California has implemented a system where track-side Board of Stewards provide independent reports on the purported cause, final reported status, and outcomes of injuries to jockeys. Other states could use the Northern California system as a model for their own horse racing injury surveillance.

The variation of regulations between states creates an additional complication for worker safety and health. Safety and health concerns could be more easily managed if regulations were more synchronized among states. This would help improve health and safety requirements and PPE use regulations.

To be competitive in this sport, jockeys must be vigilant in maintaining a low body weight. To keep a minimum weight, jockeys often resort to weight-reducing techniques, commonly known in the industry as 'wasting' and 'flipping.' These techniques pose a hazard to a jockey's long-term health. These activities may also lead to more immediate hazards if a jockey is dehydrated or otherwise not fit to ride. Representatives in other sports where weight can be an issue, like wrestling, have examined alternatives to weight requirements for keeping athletes healthy, such as minimum body fat requirements [NFHS 2006]. These alternatives should be evaluated for relevance in the horse-racing industry. Also, providing some form of health and nutritional education to jockeys would be prudent.

Many opportunities exist for research regarding worker safety and health in the horse-racing industry and injury prevention. The possibility for lead exposure should be quantitatively assessed. If it is found that the exposure limits exceed current standards, this hazard could be remedied either by using an alternative to lead weights, such as weighted pads, or by encapsulating all lead weights. An evaluation of the effect of silica or synthetic fibers on the respiratory health of workers also is needed. The potential benefits of synthetic surfaces for the well-being of the horse should be weighed against possible respiratory ailments that jockeys and horses may suffer. As with all emerging safety and health issues, NIOSH will do its best to continue monitoring the health and safety of these workers.

An effort on the part of horse-racing industry representatives (race tracks, racing commissions, and horse owners) can be taken to lessen the many hazards faced by workers in the horse-racing industry. However, the responsibility to improve the safety and health of employees in this industry lies among all participants. Below are some measures for consideration.

Industry Representatives (Race Tracks, Racing Commissions, and Horse Owners)

- Make safety and health issues a part of the everyday, decision-making processes (e.g., whether races are held, conditions for canceling a race, assessments of a jockey's fitness to ride);

- Work with jockeys and other professional associations to ensure adequate insurance and support for injured workers, while encouraging primary injury prevention practices;
- Assess the health implications of the current weight requirements and options for adjusting weights consistently in consultation with health experts;
- Develop and maintain a track-, state-, or corporate-level monitoring system to collect data on workers and their injuries and illnesses, which could serve as a model for developing a national-level surveillance system;
- Develop standards for quality on-track and off-track medical care for all facilities that include the use of staff certified in Advanced Cardiac Life Support and adequate medical equipment;
- Explore workplace and jockeys' room conditions with the intent of developing criteria for design, safety, hygiene, ventilation, and habitation;
- Integrate the safety of both humans and animals into the design of equipment and facilities (e.g., padded starting gates and safety rails);
- Support independent scientific inquiry into the dynamic health status of workers in the horse-racing industry; and
- Develop and provide appropriate education, consultation, referral, and treatment for jockeys regarding eating and weight control issues.

Jockeys
- Become educated about proper nutrition and consider healthy alternatives for weight management;
- Wear PPE and ensure that it is properly fitted and in good condition; and
- Work with industry representatives and professional associations to ensure appropriate support and follow up for injured workers, while encouraging primary injury prevention practices.

Professional Associations
- Promote the safety and health of jockeys and other race track staff by working with industry representatives;
- Work with industry representatives and jockeys to ensure appropriate support and follow up for injured workers;
- Work with industry representatives to ensure adequate on-track and off-track medical care is available at all facilities;
- Work with industry representatives to develop criteria for safe, clean jockeys' rooms; and
- Support industry representatives and jockeys in the development of appropriate education, consultation, referral, and treatment for eating and weight control issues.

Other Race Track Workers
- Become educated and trained in safety issues relevant to work responsibilities;
- Consider wearing PPE (e.g., helmets and vests) when in the vicinity of a horse; and
- Work with industry representatives and professional associations to ensure appropriate support and follow up for injured workers, while encouraging primary injury prevention practices.

References
American Horse Council Foundation [2005]. The economic impact of the horse industry on the United States. Washington DC: American Horse Council Foundation.

ASTM [2007a]. ASTM Standard F1163-04, Specifications for headgear used in horse sports and horseback riding, Annual Book of ASTM Standards, Vol. 15.07, Philadelphia, PA.

ASTM [2007b]. ASTM Standard F1937-04, Specifications for body protectors used in horse sports and horseback riding, Annual Book of ASTM Standards, Vol. 15.07, Philadelphia, PA.

Australian Racing Board [2007]. Australian rules of racing. [http://www.australian-racing.net.au/rules/rules_140607.pdf]. Date Accessed: January 2009.

Bishop K, Deans RF [1996]. Dental erosion as a consequence of voluntary regurgitation in a jockey: a case report. Br Dent J 191:343–345.

BLS [2008]. CFOI charts, 1992–2006. [http://www.bls.gov/iif/oshwc/cfoi/cfch0005.pdf]. Date Accessed: January 2009.

Burwinkle B [2002]. Jockeys: rough riders of the track. [http://www.shortsupport.org/News/0411.html]. Date Accessed: January 2009.

BETA [2007]. British Equestrian Trade Association, Body and shoulder protector standard. [http://www.beta-uk.org/Index.asp]. Date Accessed: January 2009.

California Horse Racing Board [2007]. Horse-racing rules. [http://www.chrb.ca.gov/rules_search.htm]. Date Accessed: January 2009.

California Horse Racing Board [2006a]. California Horse Racing Board rule book: Article 3.5 track safety standards: 1472. rail construction and track specifications. [http://www.chrb.ca.gov/policies_and_regulations/CalRules6-2006.pdf]. Date Accessed: January 2009.

California Horse Racing Board [2006b]. Board moves to require polytrack-like surfaces. CHRB News Release. February 16, 2006.

California Horse Racing Board [2001]. California Horse Racing Board rule book: Article 8. Running the Race: 1689.1. Safety vest required. [http://www.chrb.ca.gov/policies_and_regulations/CalRules6-2006.pdf]. Date Accessed: January 2009.

California Horse Racing Board [1978]. California Horse Racing Board rule book: Article 4 Occupational Licenses: 1501. worker's compensation insurance required. [http://www.chrb.ca.gov/policies_and_regulations/CalRules6-2006.pdf]. Date Accessed: January 2009.

CDC [2001]. Nonfatal occupational injuries and illnesses treated in hospital emergency departments—United States, 1998. MMWR 50(16):313–317.

Colton R [2007]. 2005 statistical study of jockeys' mounts, wins, and earnings. Submitted to NIOSH docket 104. [http://www.cdc.gov/niosh/docket/NIOSHdocket0104.html]. Date Accessed: January 2009.

Copeland PM [1994]. Renal failure associated with laxative abuse. Psychother Psychosom. 62(3–4):200–202.

DeBenedette V [1987]. For jockeys injuries are not a long shot. Phys Sports Med 15:237–245.

Delaware General Assembly [2007]. Administrative Code: Title 3 Agriculture: 1001 Thoroughbred Racing Commission. [http://regulations.delaware.gov/AdminCode/title3/1000/1001/index.shtml#TopOfPage]. Date Accessed: January 2009.

Delaware General Assembly [2006]. Administrative Code: Title 3 Agriculture: 1002 Delaware jockey's health and welfare board: 2.0 eligibility criteria for health coverage. [http://regulations.delaware.gov/AdminCode/title3/1000/1002/1002.shtml]. Date Accessed: January 2009.

Diamond GL, Goodrum PE, Felter SP, Ruoff WL [1998]. Gastrointestinal absorption of metals. Drug Chem Toxicol 21(2): 223–251.

Gitomer JC [2005]. Letter of August 16, 2005, from Jeffrey C. Gitomer, District Director, Employment Standards Administration, Office of Labor-Management Standards, U.S. Department of Labor, to Steve Gigliotti, Esq., Gigliotti and Gigliotti, Lawyers, attorney for the Jockeys' Guild, Inc.

Inui A, Kobayashi S, Uemoto M, Yoon S, Kasuga M [1997]. Metabolic alkalosis and chronic renal failure in a bulimic patient. Nephron. 76(3):362.

Jockeys' Guild [2007]. Jockeys sue Illinois horse owners and trainers. In: Jockeys' Guild News and Articles. July 5, 2007. [http://www.jockeysguild.com/news]. Date Accessed: January 2009.

Joint Committee on Administrative Rules [1994]. Illinois Administrative Code: Title 11: Alcohol, horse racing, and lottery: Subtitle B: Horse racing: Chapter 1: Illinois Racing Board: Subchapter g: Rules and regulations of horse racing (Thoroughbred): Part 1411 Jockeys, apprentices, jockey agents, and valets: Section 1411.240 Safety equipment. [http://www.ilga.gov/commission/jcar/admincode/011/011014110002400R.html]. Date Accessed: January 2009.

Jonat LM, Birmingham CL [2003]. Kidney stones in anorexia nervosa: a case report and review of the literature. Eat Weight Disord 8(4):332–335.

Kentucky Administrative Regulations [2007]. Title 810 Environmental and Public Protection Cabinet, Department of Public Protection, Kentucky Horse Racing Authority [http://www.lrc.state.ky.us/kar/TITLE810.HTM]. Date Accessed: January 2009.

King MA, Mezey G [1987]. Eating behavior of male racing jockeys. Psychol Med. 17:249–253.

Labadoarios D, Kotze J, Momberg D, Kotze TJ [1993]. Jockeys and their practices in South Africa. World Rev Nutr Diet *71*:97–114.

LaMarra T [2007]. Tapeta surface focus of Presque Isle meet. The Blood-Horse magazine. [http://news.bloodhorse.com/viewstory.asp?id=40851]. Date Accessed: January 2009.

Lavelle J, Murphy J [1977]. Jockey's ankle: an occupational lesion. J Ir Med Assoc *70*:282.

Leydon MA, Wall C [2002]. New Zealand jockeys' dietary habits and their potential impact on health. Int J Sport Nutr Exerc Metab *12*:220–237.

Mitchell JE, Crow S [2006]. Medical complications of anorexia nervosa and bulimia nervosa. Curr Opin Psychiatry *19*:438–443.

National Commission on State Workmen's Compensation Laws [1972]. Washington, DC: Government Printing Office.

NFHS [2006]. 2006–07 wrestling rule changes. National Federation of High Schools [http://www.nfhs.org/web/2006/08/200607_wrestling_rules_changes.aspx]. Date Accessed: January 2009.

New York State Legislature [2007]. Racing, Pari-Mutuel Wagering and Breeding Law: Article 2 Thoroughbred Racing and Breeding: 213a. [http://www.racing.state.ny.us/about/rls.home.htm]. Date Accessed: January 2009.

NIOSH [2007a]. NIOSH Docket Number 104: Horse Racing Industry. [http://www.cdc.gov/niosh/docket/NIOSH-docket0104.html]. Date Accessed: January 2009.

NIOSH [2007b]. Sixty-five-year-old jockey dies after being partially run over by mobile starting gate. In-house unpublished fatality investigation. March 2007. Morgantown, WV: Division of Safety Research.

Ohio State Racing Commission [n.d.]. Ohio Thoroughbred Racing Rules: Chapter 3769-6. Claiming, starter and jockey rules: Rule 3769-6-25.1 Safety reins, pp. 72.

Opacich KJ, Lizer S [2007]. Determining health status and health disparities for an embedded rural workforce. Submitted to NIOSH docket 104. [http://www.cdc.gov/niosh/docket/NIOSHdocket0104.html]. Date Accessed: January 2009.

Palla B, Litt IF [1988]. Medical complications of eating disorders in adolescents. Pediatrics *81*(5):613–623.

Parker M [1998]. The history of horse racing. [http://www.mrmike.com/explore/hrhist.htm]. Date Accessed: January 2009.

Pennsylvania Code [1997]. Chapter 163. Rules of racing. [http://www.pacode.com/secure/data/058/chapter163/chap163toc.html]. Date Accessed: January 2009.

Press JM, Davis PD, Wiesner SL, Heinemann A, Semik P, Addison RG [1993]. The National Jockey Injury Study: an analysis of injuries to professional horse-racing jockeys. Clin J Sports Med *5*:236–240.

Price D [1973]. Abuse of diuretics by jockeys [letter to the editor]. Br Med J. *1*(5856):804.

Racing Officials Accreditation Program [2007]. The accreditation process. [http://www.horseracingofficials.com/default.asp?id=3.1] Date Accessed: January 2009.

Tsirikos A, Papagelopoulos PJ, Giannakopoulos PN, Boscainos PJ, Zoubas AB, Kasseta M, Nikiforidis PA, Korres DS [2001]. Degenerative spondyloarthropathy of the cervical and lumbar spine in jockeys. Orthopedics *24*:560–564.

The Turf Club [2007]. Rules of racing and Irish national hunt steeplechase rules. [http://www.turfclub.ie/site/RuleBook.pdf]. Date Accessed: January 2009.

Turner L [2000]. Against all odds, medical providers keep watch as jockeys bet their health on winning rides. J Ark Med Soc *96*(10):366–369.

USA Horse Racing [2003]. USA horse racing—race tracks. [http://www.officialusa.com/stateguides/horseracing-tracks/index.htm]. Date Accessed: January 2009.

U.S. Census Bureau [2008]. Current population survey main page. [http://www.census.gov/cps/]. Date Accessed: January 2009.

Waller AE, Daniels JL, Weaver NL, Robinson P [2000]. Jockey injuries in the United States. JAMA *283*(10):1326–1328.

Washington State Legislature [2007]. Washington Administrative Code 260-12-180: Safety equipment required. [http://apps.leg.wa.gov/WAC/default.aspx?cite=260-12-180]. Date Accessed: January 2009.

West Virginia Racing Commission [2000]. Title 178 Legislative Rule Racing Commission, Series One Thoroughbred Racing. [http://www.wvracingcommission.com/rulestdf.pdf]. Date Accessed: January 2009.

NOTES

NOTES

NOTES

www.ingramcontent.com/pod-product-compliance
Lightning Source LLC
Chambersburg PA
CBHW081419170526
45166CB00010B/3399